The Scientific Life

Miguel A. Sanchez-Rey

Metamorphic Space: A Guide Through Metaspace

Miguel A. Sanchez-Rey

Table of Contents

Introduction

The history of algebraic geometry has been noticeably historical and fruitful. With advances in the sub-fields of topology and differential geometry, with the richness and scope of complex dynamical systems, algebraic geometry has been on the forefront of modern advances in homological algebra. Which has spilled over into the fields of super-symmetry and strings. What role algebraic geometry plays in the realm of modern physics has been understood to be as a guide between the universal domain and the domain of real, complex, and rational numbers which are subsets of the universal domain. Algebraic structures includes the specialization of the extended functional elements and variation of functional properties.

Homological algebra is understood to be the finite algebraic operations of covariant and Contravariant functors which exhibit both homomorphism and exactness in a commutative algebra. To understand and take advantage of the commutation of algebraic structures enables the extraction of K-theory in homology and cohomology which form direct sums and direct products.

Nevertheless one is not to be carried away by homology theory rather use homological algebra as a guide through metamorphic space. Metamorphic space being:

Cosmological homotopic states between variant [of stringy]'s.

In which homotopy is stated as two continuous functions, from one topological space to another, that can be continuously deformed into the other. That is to say variant [of stringy]'s are variant of each other which are elements of perfect number. Cosmological being the cosmological wave-function that enables comprehension of metaspace.

A Guide Through Topology and Homology

One expresses a topological space as an association of subsets that in itself constitute a topology. Constructing topological structures, whether point-set, algebraic or differential, requires that one use sets and certain finite algebra, whether product spaces, direct spaces, metric spaces or even functional spaces, to construct a viable topological space. Geometric topology being the utilization of differential forms and fibre bundles to Riemannian manifolds.

Homological algebra is said to be a commutative algebra of modules with homomorphism of particular functors, whether derived or additive, that form a homology and cohomology of augmented rings.

Algebraic Geometry

The geometry of algebraic structures states that geometric elements be included in finite algebraic operations. That is to say that functional elements be specialized and in which has a universal domain that in its subsets constitutes the real, complex and rational numbers which are variational and can be extended.

Classical and Modern Strings

Classical strings are bosonic strings that reside in a one-dimensional world-sheet. While modern strings reside in a compactified 11 - dimensional Calabi-Yau manifold, and in which, through first quantization, include fermionic and bosonic properties of harmonic variational and tangling homological structures.

Through Metamorphic Space

Define metaspace as:

Cosmological homotopic states between variant [of stringy]'s.

In which homotopic states exist in D-energy metastates.

Conclusion: Metaspace and Its Implications

Metaspace is said to carry information about the variant [of stringy]'s, of perfect number, that when accessible, leads to the harnessing of metamorphic space to achieve a terraformic reaction that with The Grand Unification Scheme allows one to take advantage of the terraformic process.

An Introduction to *Isaac's Laws*

Miguel A. Sanchez-Rey

Table of Contents

Isaac's Laws is an experiment. An experiment that ascertains and establishes the Scientific Age and what the Scientific Age means to modern society. Isaac's Laws deals with themes one dare not talk about. Themes of sexual innocence, depravity and mutilation in a world-of-ideas.

The pathology of the Post-Modernist era is analyzed. A pathology in which there is wide-spread antagonism toward the scientific process and the conventions of writing and expression. Isaac's Laws blends in archetypical elements of Post-Modernist writing and thought to ascertain the limitations of Post-Modernism.

Isaac's Laws determines that sexuality, not historical Hegelianism, plays a key role in the development and evolution of the state. Isaac's Laws conjectures a resolution to the Nazi state and delves in the difficult lessons of the holocaust and the Second World War.

Isaac's Laws is an experiment. An experiment of enormous depth and grief. A resolution to the plight of Post-Modernist insanity and mental anguish. A return to the scientific process and the restoration of historical literature.

The conclusions of Isaac's Laws are of enormous depth and complexity. Many Post-Modernist thinkers are explored and implied.

The conclusions of Isaac's Laws are that Post-Modernism led to a fail sense of high-expectations. High-expectations that the project of Post-Modernism can transcend the centrality of logo-centrism and the scientific process. Where post-structuralisms resistance to logo-centrism and the scientific method can yield a novel methodology of thought that is counter to establish pragmatics. That the diversity-of-ideas and opinions can resist oppressive power-structures.

Post-Modernism is an ill defined era. With many definitions from many different philosophers the post-modernist set to find their own definition. With some continental philosophers supposing that post-modernism leads to post-modernisms. And that post-modernism may indeed be the incalculable era but, yet again, so many explorations of its primary ideas yielded a hopeless sense of intellectual bigotry and disunity.

While French philosophers are concern with post-structuralism and intertextuality; German philosophers specialized themselves in the theory of critical interpretation. The sciences are aspects of culture and that surpassing the sciences needs the input of self-consciousness and the phenomenological reduction. The Americas reach their own conclusions favoring a pragmatist project of practicality and the experimental value of the epistemological priority.

As the West adheres to logo-centrism, and that much of planetary society resents the domination of the West, the post-modernist movement set to define ideas which are diverse in its structuralism and counter to its methodology. A conflict that incited the rise of radicalism and extremism, and the violent opposition to both statist oppression and organized resistance.

Entailing a severe identity crisis which lay the seeds of a horrific conclusion to post-modernist thought. Thought which is, nevertheless, a thought disorder that ask for a schizophrenic breakdown: one in which the sadism-of-ideas and repression led to collective mutilation and that the over-exuberance of scientific participation gave way to a senseless and perverse consciousness that desecrates sexuality and unleashes havoc to social harmony.

That of a global society that becomes confrontational with the auto-immunity of the liberal democratic state and in which the pathology of neo-fascism becomes the appealing substitute to economic depravity and the politics of the establishment.

Isaac's Laws objective is to conduct an experiment using a novel experimental apparatus. The question is what suitable apparatus is there to behold that can achieve a successful completion of such an ambitious experiment? The apparatus that is both, in its design; simplicity, and clarity, capable of delving into the collective unconscious of the modern world is that of the Wizarding World. The next question is what makes the Wizarding World such an attractive experimental apparatus? That in its design it incorporates the flaws of the post-modernist era; that in its simplicity it captures the hopes and dreams of the modern world, and that in its clarity it delves into the present issues and themes of what is at the time the end of Post-Modernism and has now become an uncontrollable scientific machine -- The Scientific Age.

In order to gain control of the scientific machine one finds resolution in such a way that harm isn't inflicted in the public sphere when the scientific process breaks down. That resolution is presented as a set of resolutions that is establish in the characters and backstory of Isaac's Laws.

Its atoms are its many characters and its measurements are its many themes and subjects. It's a high-stakes experiment in which the intensity of literature and literary imagery is use to capture vast ideas. The scope of those ideas are linear but complex enough to determine how to confront the spirit of the age.

There is no other format in which such an experiment can be conducted but in its archetype the world is confronted by the same issues as the Wizarding World. But there are many other elements where the problems-of-the-age is confrontational and unburden. But their apparatus is too limitless to have any practical use. Its only that through brevity the subconscious becomes the sublime.

Literature is the beginning of a profound form of wild-contemplation. A contemplation that sets high-standards and expectation. Sometimes high-expectations elude to the gibberish of the spoken word and the vacuum of the semantic form. Where grammar is non-empty and a fruitless standard.

The dawn of the Scientific Age seems, very much, like a warning from the distant future. A warning of humanity's futility and a reminder of the fleeting existence of a semi-advance industrial society.

Anticipation is the antagonist of the obscure. But nevertheless a requirement. How to maneuver through the maze of ideas is the task of the Poem of Sensible Obscurity. A mutilated poem that eludes to an end but nevertheless an end without its own realization or thought. A contradiction that becomes resolvable through the stream of consciousness. That of a hopeful lesson of humanity's potential for self-betterment. A pointless beginning to the experiment but a primary element to a stubborn conclusion.

Isaac's universe is a pocket watch. Time is linear and the wonders and contradictions of ideas are explored. Those ideas are central to Isaac's awakening to a small world. Yet in being very much true and exact ideas are constantly changing. One is led to believe the obvious and the certain but is unaware of the inevitability: that much of Isaac's Universe is an incomplete set of ideas that bears contradictions and resolutions; that of limitations to conventional grammar and phonetics. In which dichotomy and textuality become a necessary component to an experiment in which high-expectations is met with frequent revision to its organization and core themes.

Isaac is confronted with the inevitability of space-time: that time moves on and motion moves forward, and backwards, only that Isaac's mother pulls a trick: a trick in which Isaac knows everything but in which the only thing he knows is that she lied. Setting Isaac off to ascertain the outer-regions of his life-world where he, lacking little noticeability, confronts the existentialism of the modern world. A said existence in which the laws of physics restrict Isaac's reach and in which the human condition takes over.

Where Isaac goes is to him a return journey. Isaac asserts his right to return. Meeting only Clara out at the outskirts of her manor home he awakes to the reality of the modernist conflict. That in which a failure is an irony. Only that failure is relative to Isaac. Reason breaks-down and the counter-Enlightenment becomes more pronounce. Reason becomes unreason and desperation becomes its outcry.

Clara is burden with the prism. A prism that opens the multi-verse of surrealism and experientialism. Overreaches herself beyond the boundaries of scientific participation. In which Adrian's takes control over and disciplines her in anticipation she'll acquaint herself with Isaac. Yet learns to subdue herself in the preoccupation of youth and the serene and quiet mind-set in which she explores.

Possible Worlds

Elisha's possible worlds completes Isaac's Laws. Answers Adrian's question and finds closure in the rights of future generations.

New Atheism is fundamentalist in practice and in thought. Their core ideas of mass violent resistance against theistic fundamentalism, whether benign or extreme, unleash further violence against secular thinkers. In that sense the endurance of New Atheism's social-Darwinism is short-live.

A social-Darwinism that lost touch with the reality, and interplay, of scarcity and ideology. Causing further harm to the natural and social order. An order that has been put in place after the Nuremberg Trials.

SUPREME and Quantum Cosmology

Miguel A. Sanchez-Rey

Abstract

Assert the dominance of SUPREME using quantum topology at the cosmological scale.

November 24th, 2016

SUPREME and Quantum Cosmology

SUPREME is understood to be a homogenous topology. By homogenous topology one means a diffeormorphic mapping from the target space to the base space; vice versa [1]. That is from external control to internal control; vice versa [2].

Take a quantum topology to be a Hilbert space from Bra space to Ket space:

$< p_2 \ | 0 | \ p_1 >$ call the vacuum quantum state. A cosmological quantum topology is now understood to be a cosmological wave-function embedded in quaternion space [3].

One then demonstrates that the quantum cosmological wave-function, denoted as Ψ, can be split into individual wave-packets in which the product of such individual wave-functions is equal to Ψ. Then at the vacuum state: $\Psi = 0$. Then giving the splitting of such wave-function into its direct product of Hilbert space in quaternion space of homological algebra:

$$\Psi = \psi_i \bigoplus_{i=1}^{\infty} H$$

Then the measure [or probability amplitude] is calculated as: $\mathcal{M} = \oint \Psi$ in which $\mathcal{M} = \varnothing$ in metaspace [4, 5]. Such that $\Psi = \varnothing$ internal control and $\mathcal{M} = \varnothing$ is external control; and the inverse, is a diffeomorphic mapping.

Each region on a quantum manifold can be said to be sheave of a super-manifold. Such that:

$\Psi \in \mathbb{R}^{p|q}$ so that giving an individual topological manifold one can cut such a topological super-manifold into individual sub-manifolds and carefully rearrange them in accordance with re-arranging the individual wave-functions in quaternion space.

References

[1] Sanchez-Rey, Miguel A. SUPREME: A Homogenous Topology. Vixra.org: 2015

[2] Sanchez-Rey, Miguel A. Internal and External Control in PHPR. Vixra.org: 2015

[3] Sanchez-Rey, Miguel A. The Logical Structure of Space-Time. Vixra.org: 2011.

[4] Sanchez-Rey, Miguel A. Current Mathematical Theory in PHPR. Vixra.org: 2015.

[5] Sanchez-Rey, Miguel A. Lagrangian Operator for Electrostatic Background Field in Ω. Vixra.org: 2016.

Establishing The Second Task of PHPR

Miguel A. Sanchez-Rey

Table of Contents

Introduction

Whereas the resolution of the foreseeable catastrophic scenario, mineral depletion, is at the forefront of The First Task of the Physicalist Program [PHPR] the payoffs to its resolution are limitless. In that resolution of The First Task will yield astronomical advances in the technological and engineering sciences in the form of the interplay. Yet advances in technology: whether the milestone of PHPR [Star Gates] or the grand achievement of PHPR [Space Habitats] carries risks that inevitability pose existential risks to human survival and tranquility. Risks that entail, in the long-term, whether or not its feasible to pursue space travel or whether or not colonization of other planets [as well within our solar system] is possible.

The dangers that are pose by The First Task of PHPR, that include a terraformic fall-out or an unlikely runaway transformative reaction, due to a security breach in metaspace or a global arms race, are existential. Though without qualifications. As no other resolution presents itself that could alleviate the harm of scarcity at a planetary scale no other avenue remains present. Even then the Phoenix Project, which aims to achieve, at the same duration, interstellar travel in the form of star ships, is a fail endeavor. The only way out is through The First Task and that includes taking advantage, at a responsible level, ITER [International Thermonuclear Experimental Reactor] as a 40-year window opportunity to achieve economic sustainability and environmental recovery in which afterwards world-wide decline becomes of little notice but in which a 60-year process is set to reach full completion of The First Task of PHPR; hence a 100 year task.

Space-Habitats

Population growth, is nevertheless, the quintessential conundrum of scarcity and conflict. As economic growth continues to increase population growth begins to explode in which at a certain point a peak is reach where consumption cannot meet rising demand. Prices, in goods and services in the market economy, whether in a state-capitalist or a command economy, increase and scarcity becomes more tantamount. The lost of bio-diversity, at a national and planetary scale, becomes more visible as well. Foreshadowing conflict between warring factions, tribes, states and nations as a flight-or-fight response between starvation and greed. Causing, in end result, a breakdown to the natural order and in the transition to a wasted planet to the extreme.

The development of space-habitats is the only measure that can address rising population growth and economic need. As human survival requires, as well, the expansion from life-on-Earth to the outer-edge of the solar system to avoid any other existential risk i.e., asteroids, epidemics, solar flares, ice-age or etc. Constructing space-habitats require enormous time at the current present but with The First Task of PHPR it would only require a fraction of a fraction of that time to construct them. Vast habitats can be constructed that within a 100 year period after The First Task colonial outposts in other moons and planets [Mars] can be abandoned in favor of space-habitats as safer measures can be implemented for permanent settlement in which more and more mineral and energy resources can be gathered within the solar system for use and in which The First Task can be use to recycle those resources for reuse.

Star Gates

Yet with the resolution of mineral depletion population growth persist and so little space is left to continue on within our own solar habitat. How to move outwards into the solar system in which star-ships cannot be the possible option requires instead the development and engineering of star-gates.

To even consider the use of star-gates requires, as well, that efforts be put in place for the study and implementation of cryogenics and technological development of human rationing and learning of the psychology of long-duration space-travel that can sustain the sanity of an astronaut in a traveling space-habitat. Whereby star-gates can be constructed and the use of worm-holes can be harness to further the development of an intergalactic transit system that shortens the time frame with near light-speed travel [for example: photonic laser thrusters or even anti-particle engines] being use to facilitate the construction of such transit system.

Within two centuries after The First Task is completed a vast intergalactic transit system will be realized in which interstellar travel becomes second nature to human civilization on planet Earth and beyond.

To move beyond the solar system risks encroaching into the solar habitats of other intelligent and primitive beings. Life, as we know it, within our own planet, is diverse in which gene selection is the reigning Darwinian theory.

In that manner within a time-span of 2 million years' mammals have become the dominant species in which predatory behavior remains the most selective attribute for survival and the development of intelligence. Humans, by and large, remain the only species capable of intelligent language and the capacity for conscious awareness.

It's expected that the Milky Way is teeming with life but very little of that life is of intelligent life capable of advance mathematics and technology. Yet even then intelligent life is scattered within our Milky Way very much as intelligent life was scattered on Earth in the metaphor of the division between island chains and the division of continents that led to phenotypical gene diversity in the human population.

Yet small incremental differences in D.N.A. or environment in any planetary system [whether or not Earth-like] make outside solar-system colonization of planetary bodies highly unlikely. And that encroaching into their solar habitats may spark an unintended conflict. And so then, only in the rare occasion, the use of The First Task of PHPR is better suited for the development of a feasible planetary habitat and on most occasions the construction of space-habitats. Yet again minerals can be preserve but more minerals are needed.

As such the risk is pose that extraction of minerals will require utilizing asteroids, stars, and etc., for the sustainability and tranquility of human life thus posing severe risk of inciting/starting a conflict in which there is no going back and no end until the threat of conflict

is neutralized or extinguished.

Defensive Measures Through Metaspace

A more likely scenario of a conflict between an extraterrestrial being is an extraterrestrial encounter gone wrong. In which in a matter of a week the first target of destruction is the armada's and space-habitats at the outer and inner edge of the solar system. The second target is the planetary colonies and the third, and last target, is planet Earth in which a five-week window of opportunity is giving to take preventive measures with metaspace.

That five week of window of opportunity is to be use for evacuation of planet Earth through the use of star-gates and the self-destruction of planet Earth is sought-out for successful evacuation and resettlement. If no evacuation and resettlement is sought-out than an advance extraterrestrial intelligence will overwhelm all defensive measures of planet Earth and the terraformic process will be use, by such beings, for its destruction since no habitation is possible. Domination is not probable since stalemate is apparent and in which prolong stalemate becomes exhaustive to human survival against a more advance extraterrestrial intelligence in which the human capacity cannot withstand.

The Second Task

The Second Task of PHPR is establish as the resolution to an extraterrestrial conflict. In which the question is pose, giving the diversity of primitive and intelligent life in our Milky Way, how does one proceed carefully beyond Earth's solar system without becoming an existential threat to an extraterrestrial intelligence?

An extraterrestrial conflict poses overwhelming problems and risks to human tranquility, longevity and survival. Its resolution will take the efforts of biologists, chemists, physicists, and engineers to make the safe-transition to the awareness and revolution of an extraterrestrial encounter and even the initiation of peaceful First Contact in which man-kind will, in its finality, have universal conception of intelligent life-beyond Earth.

It's then that awareness must be brought that, in all likely-hood, intelligent life chooses isolation from other beings beyond their space-habitats or solar habitats and in such that choice must be respected. But one must not normalize negligence as the care of these beings are also of upmost importance that they may not pose an existential threat to human beings in the long-term and in which peaceful co-existence is preferable and internationalism becomes the model.

Conclusion: Advances in the Biological Sciences and in Medicine

The resolution to an extraterrestrial conflict will yield a paradigm shift in the field of the

biological sciences and in the evolution of medicine. Whereby human health is preserved and

human survival is extended. Thereby foreshadowing the choice to pursue biological

immortality through integration of machine technology to human biology in the formality of

cybernetics and advance artificial intelligence.

SUSY-Like Electrostatic Background Fields in Metaspace

Miguel A. Sanchez-Rey

Abstract

Demonstration of SUSY [super-symmetry]-like electrostatic backgrounds fields of quantum differential topological K-theory.

December 6th, 2016.

Introduction to SUSY-like background fields in metaspace demonstrates a cohomological approach. A cohomological approach that is the dominant mathematics of super-symmetry. There background fields in metaspace point to echoes that are signatures of electrostatic dynamics. Where metaspace is cosmological homotopic states between variant [of stringy]'s of prime. Homotopic states can then be stated to exist in a parameterized subspace that is a cohomological mapping. There K-theory is very much shown to be the dominant mathematics in metaspace where SUSY-like physics is tantamount but in which there is parameterization of metaspace such that there is no degeneration of the background fields so that the background fields is shown to be SUSY-like quaternion manifolds of equivalent differential topology in quantum space-time with background signature [+ , - , - , - , ... , -] at 11-D metaspace. 11-D metaspace which is distinguished from 11-D hyperspace in first quantized conformal string topology. As well within metaspace the SUSY-like quaternion manifolds can be said to exist in D-4 variant [of stringy] metaspace -- such that D_0- variant [of stringy]'s is bosonic-like strings that reside in { modulo $1 + 2 \leq D \leq$ modulo $2 + 2$ [2 +1] } metaspace and in which D_4- variant [of stringy]'s is both bosonic-like and fermionic-like strings that reside in { modulo $0 \leq D \leq$ modulo $1 + 1$ } metaspace, that don't include both tachyon-like and dilaton-like particles, of prime.

11-D metaspace is distinguishable from 11-D supergravity in that the quaternion manifold is not to be seen as a curled in Kalabi-Yau manifold but as a D-variant manifold subspace parameterized by The Prime Factorization of PHPR so that there is no endless metamorphic states in D-energy metastates in which computational control and SUPREME is impose to bring order into metaspace.